Name: Eng. / Mostafa Yacoub Abdellatif Mahmoud

Nationality: Egyptian

ORCID: 0000-0002-9991-4624

Qualification: civil engineer Cairo University 2003

## The summation of all composite numbers:

In this paper, we will discover a formula for the summation of all composite numbers based on my discovered formula that connects prime numbers to composite numbers that belong to the Array of odd numbers (Array PTBP) Within ranges that extend to infinity.

- **My discovered formula:**

**Definitions:**

Array PTBP

It is the following Array of odd numbers

$$\begin{vmatrix} 1 & 3 & 7 & 9 \\ 11 & 13 & 17 & 19 \\ 21 & 23 & 27 & 29 \\ 31 & 33 & 37 & 39 \\ 41 & 43 & 47 & 49 \\ 51 & 53 & 57 & 59 \end{vmatrix}$$

And so on....

- For a given set of consecutive primes whose numbers =n that start with prime 3 and end with prime F and not including prime 2 and prime 5

i.e.

set=[3,7,11,13,……………………………………

………………………,F]

S=product of those consecutive primes

i.e

$$S = \prod_{i=3}^{i=F} (i)$$

Range=$R_k$ = 10 × S × k

Where  k=[1,  2,  3,  4, ................., ∞(infinity)

i.e $R_1$=10 x S x 1 and  $R_2$=10 x S x 2

And so on

- **Number of composite numbers that belong to Array PTBP and created by** the effect of those consecutive primes within the range $R_K$

- $=[(K \times 4^{\times\frac{S}{3}}) + ( \sum_{j=7}^{j=F}(K \times 4 \times (\frac{S}{j}) \times \prod_{i=7}^{i=prime\ number\ before\ current\ prime\ number\ j}(\frac{i-1}{i}))] - (n)$

Where j =consecutive values of primes

7, 11, 13,.............., F

And i= consecutive values of primes 3, 7, 11, 13,........, prime before current j prime

- The previous formula can be applied for any number of consecutive prime numbers that start with prime number 3

- The first term $(k \times 4 \times \frac{S}{3})$ represents the count of unique Composite numbers +1 that belong to the Array PTBP and are created by prime number 3 within the range.

$$R_k = 10 \times S \times k$$

- The second term

$$\sum_{j=7}^{j=F}(K \times 4 \times (\frac{S}{j}) \times \quad i = prime\ number\ before\ current\ prime\ number\ j \quad \prod_{i=7} (\frac{i-1}{i})$$

Represent the count of unique Composite numbers+n-1 that belong to the Array PTBP and are created by each prime number after the prime number 3 within the range $R_k = 10 \times S \times k$

- The third term (-n)
  Subtracting n (number of consecutive primes starting from prime number 3) because the count of composite numbers generated from those

consecutive primes includes the count of those primes in the range $R_k = 10 \times S \times k$

- **Explanation and proof for my theory in my previous paper (prime number theory)**
- **We will mention only the concept of the number cycle**

We can use the number cycle concept to understand the behavior of consecutive primes in creating composite numbers.

i.e.

$$S = \prod_{i=3}^{i=F} (i)$$

**Range=cycle range= $R_k$ = 10 × S × k**

Where k= [1, 2, 3, 4, ................., ∞(infinity)

i.e. $R_1 = 10 \times S \times 1$ and $R_2 = 10 \times S \times 2$

And so on

- Now consider only one k value =1 And Now
- For any set of consecutive primes

i.e.

set=[3,7,11,13,.............................
.........................,F]

S=product of those consecutive primes

i.e

$$S = \prod_{i=3}^{i=F} (i)$$

Range=$R_k$ = 10 × S × k

- Any cycle containing two types of numbers

  First type (numbers that are divisible by prime numbers within the set).

The second type (numbers that are not divisible by prime numbers within the set).

- For example, the following figure (considering the set of consecutive primes=[3,7]) shows the two types of numbers the colored one represents the first type while the uncolored numbers represent the second type

- We can see there must be a repeated pattern for each type of number and for both of them together each cycle up to infinity i.e for

k= [1, 2, 3, 4, ................, ∞(infinity)

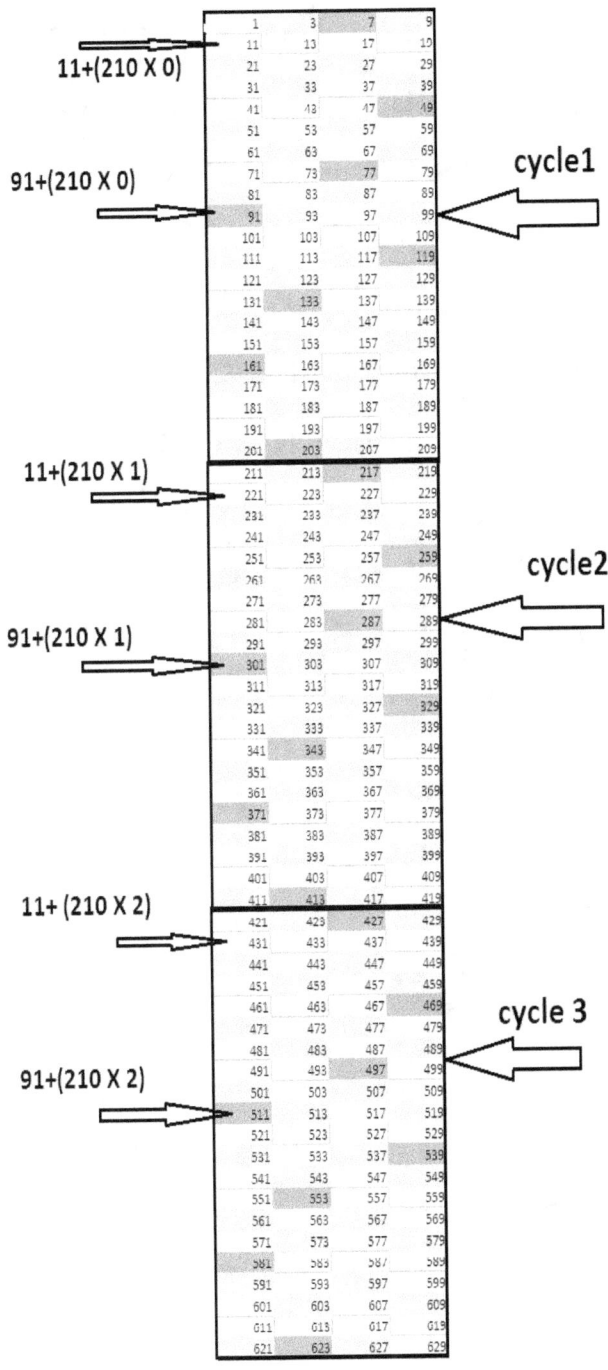

We have proved ( in my paper prime number theory part3) that the Summation of all numbers within Range = $S \times 10$

Including all numbers of the array PTBP (with last digit= 1 or 3 or 7 or 9) and all other even or odd numbers that are divisible by two or five (with last digit = 0 or 2 or 4 or 6 or 8 or 5)

$$=[[\ 50\ \times\ (\prod_{i=3}^{i=F}(i \times i)\ )\ ] + [5 \times \prod_{i=3}^{i=F}(i)\ ]]$$

Example 1

For the set of consecutive prime numbers = [3]

The summation of all numbers that belong to the following array within the range

= 10 x 3 = 30

```
 1   2   3   4   5   6   7   8   9  10
11  12  13  14  15  16  17  18  19  20
21  22  23  24  25  26  27  28  29  30
```

=[50 x 3 x 3] + [5 x 3] = 450 + 15 = 465

Now if the set of consecutive prime numbers contains all numbers with the last digit 1,3,7,9

i.e

set=

[3,7,9,11,13,17,………………….…..

Let

$$\sum \text{all prime numbers} = \text{the summation of all prime numbers}$$

And let

$$\sum \text{all composite numbers} = \text{the summation of all composite numbers}$$

$$[\sum \text{all prime numbers}] + [\sum \text{all composite numbers}]$$

$$= [[\ 50\ x\ (\prod_{i=3}(i\ x\ i)\ )] + [5\ x\ \prod_{i=3}(i)\ ]]$$

**Where $i$ takes values 3,7,11,13,………………….**

i.e

$$\sum \text{all composite numbers} =$$

$$\left[\left[50 \times \left(\prod_{i=3}(i \times i)\right)\right] + \left[5 \times \prod_{i=3}(i)\right]\right]$$

$$-\left[\sum \text{all prime numbers}\right]$$

**Where**

$\sum \text{all prime numbers} = \sum P$ the summation of all prime numbers

And

$\sum \text{all composite numbers} = \sum C$ the summation of all composite numbers

And

$$\left[\frac{\sum P}{\sum C}\right] = \left(\frac{1}{\dfrac{50x\left(\prod_{i=3}(i \times i)\right) + 5x\prod_{i=3}(i)}{\sum P}} - 1\right)$$

**This formula explains how rare prime numbers are**

.

www.ingramcontent.com/pod-product-compliance
Lightning Source LLC
Chambersburg PA
CBHW072058230526
45479CB00010B/1134